YOUR KNOWLEDGE HAS VALUE

- We will publish your bachelor's and
  master's thesis, essays and papers

- Your own eBook and book -
  sold worldwide in all relevant shops

- Earn money with each sale

Upload your text at www.GRIN.com
and publish for free

Urgessa Tilahun

Farmers' Perceptions and Adaptations to Climate Change through Conservation Agriculture

The Case of Guto Gida and Sasiga Districts, Western Ethiopia

GRIN Verlag

Bibliografische Information der Deutschen Nationalbibliothek:

Die Deutsche Bibliothek verzeichnet diese Publikation in der Deutschen National-
bibliografie; detaillierte bibliografische Daten sind im Internet über http://dnb.d-
nb.de/ abrufbar.

Imprint:

Copyright © 2013 GRIN Verlag GmbH
Druck und Bindung: Books on Demand GmbH, Norderstedt Germany
ISBN: 978-3-656-71160-5

This book at GRIN:

http://www.grin.com/en/e-book/277938/farmers-perceptions-and-adaptations-to-
climate-change-through-conservation

Farmers' Perceptions and Adaptations to Climate Change through Conservation Agriculture: The Case of Guto Gida and Sasiga Districts, Western Ethiopia

Urgessa Tilahun[1]

[1] Haro Sabu Agricultural Research Center, Ethiopia

i.   Acknowledgement

Many thanks and appreciation goes to the following institutions and individuals whom without their help and support, the successful completion of my study would not have been possible. I am very grateful to my advisors for their guidance and encouragement to accomplish this work. I am also highly indebted to Ato Abera Gemechu, Socio-economic Researcher at Debre zeit Agricultural Research Center for his support in supplying me with necessary related journals and articles.

ACCCA   Advancing Capacity to Support Climate Change Adaptation

CA   Conservation Agriculture

km   Kilometer

OSSREA   Organization for Social Science Research in Eastern and Southern Africa

SPSS   Statistical Package for Social Science

VIF   Variance Inflation Factor

iii.   Table of Contents

Table of Contents

v.   Abstract

Ethiopia, one of the developing countries, is facing serious natural resource degradation problems. The main objective of this study was to examine the farmer's perceptions and adaptation to climate change through conservation agriculture. The data used for the study were collected from 142 farm households heads drawn from five kebeles. Primary data and secondary data were used. In addition to descriptive statistics, Heckman two stage sample selection model was employed to examine farmer's perceptions and adaptations of climate change. Farmers level of education, household nonfarm income, livestock ownership, extension on crop and livestock, households' credit accessibility, perception of increase in temperature and perception of decrease in precipitation significantly affect the adaptation to climate change. Similarly, farmers' perception of climate change was affected significantly by information on climate, farmer to farmer extension, local agro -ecology, number of relatives in development group and perception of change in duration of season. A binary logit model was employed for farmers' participation in conservation agriculture shows education level, number of active family labour and main employment of farmers were significant variables in determining participation in conservation agriculture

Key words: Climate Change, Conservation Agriculture, Heckman and Binary Logit, Western Ethiopia

1. Introduction

Human beings of current world are faced by the depletion of natural resource (Abera, 2003). Agriculture is among the factors affecting the environment in satisfying human needs, while "climate is the primary determinant of agricultural productivity" (Apata *et al.*, 2009).

Ethiopia, one of the developing countries, is "facing serious natural resource degradation problems" (Anemut, 2006). The diversity in altitude accompanied with climatic and ecological variations which affect production is among the features of the country (Shibru & Kifle, 1998). One of Ethiopia's principal natural resources is its rich endowment of agricultural land. Agriculture is the backbone of the Ethiopian economy and is given special attention by the government to spearhead the economic transformation of the country. However, land degradation, especially soil erosion, soil nutrient depletion and soil moisture stress, is a major problem confronting Ethiopia. The proximate causes of land degradation include cultivation of steep slopes and erodible soils, low vegetation cover of the soil, burning of dung and crop residues, declining fallow periods, and limited application of organic or inorganic fertilizers.

Climate is a primary determinant of agricultural productivity. The rate and magnitude of change in climate characteristics determines agronomic and economic impacts from climate change (Bruce *et al.*, 2001). Though climate change is a threat to agriculture and non-agricultural socio-economic development, "agricultural production activities are generally more vulnerable to climate change than other sectors" (Ayanwuyi *et al.*, 2010).

Literature on farmers' perceptions about climate change and participation on conservation agriculture in Ethiopia in general and in the Oromia Region in particular are very few. There are no empirical studies conducted on farmers' perceptions of climate change and their adoption decision on agricultural conservation strategies in Guto Gida and Sasiga districts.

The purpose of this study is therefore, to examine the farmers' perceptions and adaptation to climate change through conservation agriculture in which the following specific objectives, examine farmers' perceptions and adaptations to climate change, investigate farmers' perception towards conservation agriculture as adaptation strategy to climate change and analyze the determinants of farmers' participation in conservation agriculture, were studied.

2. Materials and Methods

This paper used both primary and secondary data. Primary data was collected by structured questionnaire. Detailed information on household and farm characteristics, household socio-economic and demographic characteristics, location characteristics and farm management practices and other related information were collected through interview of sample household heads.

The study was conducted in Guto Gida and Sasiga districts, East Wollega Zone of Oromia Regional State. These districts were purposefully selected due to the fact that in these areas the environment has been degraded largely and the occurrence of climate change that affect agricultural production during the year 2010 and 2011 in three kebeles of Guto Gida district. Systematic random sampling technique was employed to draw sample of household heads. From a total of 50 peasant associations in these districts nine peasant associations were selected randomly. From these sampled peasant associations based on formula by Kothari (2004) 142 households were selected proportionally.

Two types of econometric models were used for this study. The first model, Heckman Two Stage Selection Model, analyzes farmers' perception and adaptation to climate change, whereas the second model, Binary Logistic Regression Model, examines the farmers' participation in conservation agriculture in Guto Gida and Sasiga districts of Oromiya Regional State.

Statistical Package for Social Science (SPSS) version 16.0 and stata version 10.0 were employed for the analysis of this study. Along with the econometric models, descriptive statistics tools were employed to have clear picture of household demographic characteristics, socio-economic and farm characteristics, perception and adaptation of climate change and participation in conservation agriculture. Mean, standard deviation, percentage, t-test, χ^2 test, Wald test, correlation matrix and charts were employed to analyze data.

Adaptation to climate change involves a two-stage process: first, perceiving change and, second, deciding whether or not to adapt by taking a particular measure. This leads to a sample selectivity problem, since only those who perceive climate change will adapt, whereas we need to make an inference about adaptation by the agricultural population in general, which implies the use of Heckman's sample selectivity probit model (Maddison, 2006). The probit model for sample selection assumes that an underlying relationship exists, the latent equation given by

$$y_j^* = x_j \beta + u_{1j} \quad \text{-- (1)}$$

such that we observe only the binary outcome given by the probit model as

$$y_j^{probit} = \left(y_j^* > 0 \right) \quad \text{-- (2)}$$

The dependent variable is observed only if j is observed in the selection equation

$$y_j^{select} = \left(z_j \delta + u_{2j} > 0 \right) \text{-- (3)}$$

$$u_1 \sim N(0,1)$$

$$u_2 \sim N(0,1)$$

$$corr(u_1, u_2) = \rho$$

Where x is a k-vector of regressors or independent variables that is affect farmers perception and adaptation to climate change, z is an m vector of regressors, u_1 and u_2 are error terms. When $\rho \neq 0$, the standard probit techniques applied to equation (1) yield biased results (Deressa *et al.*, 2008). Thus, the Heckman probit provides consistent, asymptotically efficient estimates for all parameters in such models. Thus, the Heckman two stage selection model was employed to analyze the perception and adaptation to climate change in the Guto Gida and Sasiga districts.

For this study, the first stage of the Heckman probit model considers whether the farmer perceived a climate change; this is the selection model. The second-stage model looks at whether the farmer tried to adapt to climate change, and it is conditional on the first stage, that is, a perceived change in climate. This second stage is the outcome model (Deressa *et al.*, 2008).

There are two dependent variables; farmers' perception of climate change and farmers' adaptation to climate change. Farmers' perception of climate Change (climate_perception) is selection equation and dichotomous in nature and represented in the model 1 for perceived farmer, otherwise 0. Farmers' adaptation to climate change (climate_adaptation) is outcome equation and dichotomous in nature and explains whether farmers adapted climate change or not. It is valued 1 in the model if farmer adapted climate change, 0 otherwise. The explanatory variables for the selection equation include different socio-demographic and environmental factors based on the literature on factors affecting the awareness of farmers to climate change or their risk perceptions. The explanatory variables of the outcome equation are chosen based on the climate change adaptation literature and data availability. These variables include: education of the head of the household, household size, gender of the head of the household, non-farm income, livestock ownership, extension on crop and livestock production, access to credit, farm size, distance to input and output markets, temperature and precipitation.

A logistic regression analysis was employed to identify the factors that influence farmer's participation in conservation agriculture as an adaptation to climate change. The farmers' participation in conservation agriculture is dependent variable which takes a value of 1 if the farmer was participated and 0 if farmer did not participated. The basic model of the logit estimation (Gujarati, 2004) is as follows:

$$p_i = prob(y_i = 1) = \frac{1}{1 + e^{-(\beta_0 + \beta_1 x_{1i} + \ldots + \beta_k x_{ki})}}$$

$$= \frac{e^{(\beta_0 + \beta_1 x_{1i} + \ldots \pm \beta_k x_{ki})}}{1 + e^{-(\beta_0 + \beta_1 x_{1i} + \ldots \pm \beta_k x_{ki})}} \quad \ldots\ldots\ldots\ldots\ldots\ldots\ldots\ldots\ldots\ldots \quad (4)$$

Similarly,

$$p_i = prob(Y_i = 0) = 1 - prob(Y_i = 1)$$

$$= \frac{1}{1 + e^{(\beta_0 + \beta_1 x_{1i} + \ldots \pm \beta_k x_{ki})}} \quad \ldots\ldots\ldots\ldots\ldots \quad (5)$$

By dividing (4) by (5) we get

$$\frac{\mathbf{Prob}(Y_i = 1)}{\mathbf{Prob}(Y_i = 0)} = \frac{p_i}{1 - p_i} = e^{(\beta_0 + \beta_1 x_{1i} + \ldots \pm \beta_k x_{ki})} \quad \ldots\ldots\ldots\ldots\ldots \quad (6)$$

Where P_i is the probability that household participate in conservation agriculture and then ($1-P_i$) is the probability that household is non participant in conservation agriculture and e is the exponential constant.

The two computing models commonly used in the adoption studies are the probit and logit models. But the results obtained from the two models are very similar since the normal and logistic distributions from which the models are derived are very similar (Gujarati, 2004). As a result, only the logit model will be reported in the paper even if both models will be estimated for the purpose of comparison.

In this analysis before estimating the model, it was necessary to check the existence of *multicollinearity* among the hypothesized explanatory variables. *Multicollinearity* problem arises when at least one of the independent variables is a linear combination of the others; with the rest that we have too few independent normal equations and, hence, cannot derive estimators for our entire coefficient. VIF shows how the variance of an estimator is inflated by the presence of *multicollinearity* (Gujarati, 2004). The speed with which variances and covariances increase can be seen with the variance-inflating factor (VIF) , which is defined as $VIF_j = \dfrac{1}{1 - R_j^2}$ where

R_j^2 is the coefficient of determination in the regression. The larger the value of VIF$_j$, the more troublesome or collinear the explanatory variables is (Gujarati, 2004).

Farmers' participation in conservation agriculture (Participation_CA), for logit analysis has a dichotomous nature measuring the willingness of a farmer to participate in conservation agriculture as a measure of adaptation of climate change. The probability of participation in conservation agriculture practices dependent on several household, farm and location characteristics. The independent variables included in this model were age, sex, marital, total family size, level of education, topography of arable land, farming experience, farm size in hectares, extension services and technology promoters, membership in farmer organization, main employment, and active family labor.

3. Result and Discussion

From all sampled respondents 69 were taken from Guto Gida and the left 73 were sampled from Sasiga district. Out of all these 109 respondents perceived the change in climate while the remaining 33 did not perceived the change in climate (Appendix a). Farmers who perceived change in climate have around 8 mean numbers of relatives of household head in development group while it was around 6 for those who did not perceive the change in climate. The maximum number of relatives of respondent household heads who did not perceive climate change was 24 while it was 23 for those who perceived the change (Appendix a). The t-test values indicated that the difference in number of relatives of households in development group between those who did not perceive the change in climate and those who perceived the climate change was significant at 1 percent probability level (Table 1). The average farm income during last production period (2012/13) for the household those who did not perceived the change in climate was 4,126.57 and the mean of farm income of those who perceived the change in climate was 8,909.20. The t-test values indicated that the difference in farm income between those who did not perceive the change in climate and those who perceived the climate change was significant at 1 percent probability level (Table 1).

The mean of nonfarm income during last production period (2012/13) for farmers who did not perceive and who perceived change in climate was 2,930.30 and 4,380.96 respectively. The t-test values indicated that the difference in nonfarm income of households between those who did not perceive the change in climate and those who perceived the climate change was significant at less than10 percent probability level (Table1).

Table 1. Summary statistics of continuous variables and their mean difference test used in selection equation for the Heckman two stage selection model (n=142)

List of variables	Total respondent		Not perceived[1]		Perceived[2]		t -Value
	Mean	St. d	Mean	St. d	Mean	St. d	
No_of_relatives	7.4155	4.4867	5.61	4.795	7.96	4.262	2.703***
Farm Income	7797.75	5695.92	4126.57	2777.30	8909.20	5891.21	4.506***
Non-Farm Income	4043.83	3813.69	2930.30	3397.16	4380.96	3882.68	1.933*

***, ** and * significant at 1%, 5% and 10% respectively

Source: Own Survey, 2013

The maximum level of education of the respondent household who did not perceive change in climate was those attained grade 9-10 while the maximum level of education for household head who perceived the occurrence of climate change was those with certificate. Out of all households who perceive the occurrence of the climate change 39.45 percent were those who attended grade 1-8 (Appendix a). The χ^2 test shows significant difference between households who perceived the climate change to those who did not perceive the change (Table 2).

Change in duration of season was perceived differently among the respondent households in the study area. The χ^2 statistic (11.636) and its small significance level (p< .001) indicate that it is very unlikely that these variables are independent of each other. This shows the existence of relationship between a household's perception of climate change and their perception in change in duration of season (Table 2).

Having information on climate change is one way through which farmers perceive the change in climate. Variability in accessibility of information on climate change between those who did not perceived the change in climate and those who did was the same The χ^2 statistic (56.119) and its small significance level (p< .001) indicates existence of relationship between a household's perception of climate change and their availability of information on climate change (Table 2).

Farmer to farmer extension helps the farmers to share experience and information between them in perceiving environmental problems occurring in their area. The χ^2 test shows significant

---

[1] Farmers who did not perceive climate change
[2] Farmers who perceived the change in climate

difference between households who received farmer to farmer extension to those who did not take the extension (Table 2).

Table 2 Summary statistics of dummy and categorical variables used in selection equation for the Heckman two stage selection model (n=142)

List of variables	Total respondent		Not perceived[3]		Perceived[4]		χ^2 -Value
	Mean	St. d	Mean	St. d	Mean	St. d	
Education	1.2887	1.1397	0.79	1.023	1.44	1.134	13.353**
Season_change	0.4718	0.5009	0.21	0.415	0.55	0.50	11.636***
Information	0.6831	0.4669	0.15	0.364	0.84	0.364	56.119***
Farmer_extension	0.7253	0.4479	0.27	0.452	0.86	0.346	44.211***
Local Agroeco	0.4507	0.4993	0.61	0.496	0.40	0.493	4.191**

***, ** and * significant at 1%, 5% and 10% respectively

Source: Own Survey, 2013

Perceiving climate change is prerequisite for adaptation of climate change. Out of total of 109 respondents who perceived the change in climate was 75 respondents adapted the change through taking adaptation measures while 34 of from 109 respondents did not adapt the change (Appendix a).

The maximum family size for household head those who did not adapt the change in climate was 9 and the minimum family size was 2 (Appendix b). The average family size of those who did not adapt to climate change was around 5 and the family sizes of the household head those who did not adapt the change deviates from its mean by 1.805. However, the maximum family sizes of respondent household those who perceive the change in climate was 16 while the minimum was 3. The standard deviation of family size of those farmers who adapt to climate change was 2.147. This shows that the family size of respondents who did adapt the change in climate deviates larger from its mean than those who did not adapted the change in climate. The t-test values indicated that the difference in family size of households between those who did not adapt the change in climate and those who adapted the climate change was significant at 1 percent probability level (Table 3). The maximum farm size for those farmers who did not adapt the

---

[3] Farmers who did not perceive climate change
[4] Farmers who perceived the change in climate

change in climate was 3 hectare while it was 7.25 hectare for those who adapted the change in climate (Appendix b). As the result of the survey shows the mean farm size of respondents who adapted the change in climate was 1.537 hectare which is greater than mean farm size of respondents who did not adapt the change in climate.

The mean of nonfarm income for farmers who did not adapt and who adapted the change in climate was 2,132.740 and 5,400 respectively. The standard deviation of the household nonfarm income for farmers who did not adapt the change in climate was 1,871 and 4,131 for farmers who adapted the change in climate. The t-test values indicated that the difference in nonfarm income of households between those who did not adapt the change in climate and those who adapted the climate change was significant at 1 percent probability level (Table 3). The mean distance from input market for those who did not adapt the change in climate was 12.519 km while it was 15.195 km for those households who adapted the change in climate. The standard deviation of the respondent households distance from input market was 10.36 for those who did not adapt the change in climate and 11.88 for those who adapt the change in climate change. The mean distance from output market for those who did not adapt the change in climate was 10.61 km while it was 15.12 km for those households who adapted the change in climate. The standard deviation of the respondent households distance from output market was 10.16 for those who did not adapt the change in climate and 14.04 for those who adapt the change in climate.

Table 3 Summary statistics of continuous variables and their mean difference test used in outcome equation for the Heckman two stage selection model (n=142)

List of variables	Total respondent		Not adapted[5]		Adapted[6]		t -Value
	Mean	St. d	Mean	St. d	Mean	St. d	
Family_Size	6.183	2.220	4.88	1.805	6.77	2.147	4.467***
Farm_Size	1.442	1.007	1.2338	.81823	1.5370	1.07327	1.464
Non-Farm Income	4380.96	3882.68	2132.74	1871.07	5400.16	4131.02	4.403***
Distance_Input	14.36	11.45	12.52	10.36	15.19	11.88	1.132
Distance_ Output	13.71	13.08	10.61	10.16	15.12	14.04	1.684**

*** and ** significant at 1% and 5% respectively

Source: Own Survey, 2013

---

[5] Households who did not adapt climate change
[6] Households who adapted climate change

Farmers those who adapted the change in climate were 75 out of which 6.6 percent were female households while those who did not adapt were 34 out of which 17.65 percent were female headed households. The maximum level of education of the respondent household who did not adapt climate change was those households who have certificate while the maximum level of education for household head who adapted the occurrence of climate change was those households who attained grade 11-12. The standard deviation of education level of household adapted change in climate was 0.97 while it was 1.163 for those farmers who did not adapt change in climate. This shows that variability of level of education of households was larger for those who did not adapt the change in climate than those who adapted the change. Out of all households who adapted the occurrence of the climate change 53.33 percent were those who attended grade 1-8 while 58.82 percent (20 out of 34) of all who did not adapt climate change were those who were illiterate (Appendix a). This implies that illiterate households have more probability not to adapt climate change than those with higher level of education.

 Extension on crop and livestock is one way through which households exchange information to each other. Out of 109 household heads those perceived climate change 74 of them were those who get extension on crop and livestock. From the total of 109 farmers who did not adapt and adapted climate change 74 were those who received extension on crop and livestock and 35 of them were those who did not receive the extension.

The availability of credit may facilitate the favorable condition to adapt climate. As per the result of household survey reflected 57.04 percent of the total households were those with no availability of credit. The χ^2 statistic (11.855) and its small significance level (p< .001) indicate existence of relationship between a household's who with access to credit and those without the access (Table 4). From total of 69 respondents from Guto Gida 55 households perceived climate change. Out of these who perceive the change in climate 51 of them perceive increase in temperature and three of them respond as there is no change in temperature. From total 54 farmers in Sasiga district who perceive change in climate 42 of them respond as temperature is increasing, 6 of them perceived decrease in temperature and the left 6 farmers responded as there was no change in temperature (Appendix c).

From all the respondents on perception of change in precipitation 41 respondents from Guto Gida and 49 respondents from Sasiga districts were those who perceive decrease in precipitation. 9 respondent from Guto Gida and 2 from Sasiga district were perceived increase in precipitation while the rest 5 from Guto Gida and 3 from Sasiga district were those who did not observe change in precipitation (Appendix c).

Table 4 Summary statistics of dummy and categorical variables used in outcome equation for the Heckman two stage selection model (n=142)

List of variables	Total respondent		Not adapted[7]		Adapted[8]		χ^2 -Value
	Mean	St. d	Mean	St. d	Mean	St. d	
Education	1.440	1.134	0.74	1.163	1.76	0.970	32.244***
Sex	0.899	0.303	0.82	0.387	0.93	0.251	3.109*
Livestock_Ownership	0.789	0.409	0.41	0.500	0.96	0.197	42.235***
Extension_on_crop	0.678	0.469	0.29	0.462	0.85	0.356	35.560***
Credit	0.449	0.499	0.21	0.410	0.56	0.500	11.855***
Increase_temperature	0.853	0.355	0.529	0.506	1	0	41.366***
Decrease_precipitation	0,825	0.381	0.441	0.504	1	0	50.760***

*** and * significant at 1% and 10% respectively

Source: Own Survey, 2013

Conservation agriculture is one of the mechanisms of climate change adaptation. This study was also conducted in above stated districts in which 142 respondents were interviewed to know their participation in conservation agriculture (Appendix a).

The average age of sample household heads for those who did not participate on conservation agriculture was 38.21 with standard deviation of 12.55. The mean age of respondents who participated on conservation agriculture was 48.58 and the age of respondents who participate on conservation agriculture was deviates from its mean by 13.73. The minimum age of the respondent households was 22 and the maximum age of the respondent was 90 (Appendix d). The maximum farm size for those farmers who did not participate on conservation agriculture was 7.250 hectare while it was 4.75 hectare for those who participated on conservation agriculture. As the result of the survey shows the mean farm size of respondents who participated on conservation agriculture was 1.364 hectare which is greater than mean farm size of respondents who did not participate on conservation agriculture which is 1.332 hectare.

---

[7] Households who did not adapt climate change
[8] Households who adapted climate change

The mean years of farming experience of respondent households who did not participate on conservation agriculture was much less than those who participated on conservation agriculture. The t-test values indicated that the farming experience between those who did not participate on conservation agriculture and those who participated on conservation agriculture was significant at 1 percent probability level (Table 5). This shows that farmers with high years of experience highly participate on conservation agriculture than farmers with less years of experience.

The maximum active family labor for respondent household was 13. The mean of active family labor of households, those who participated on conservation agriculture (4.98) was higher than those who did not participate on conservation agriculture which was 2.70. This shows that the size of active family labor in households family size affect participation on conservation agriculture.

The maximum family size for household head those who did not participate on conservation agriculture was 12 and the minimum family size was 2. The mean family size of those who did not participate on conservation agriculture was 5.1 and the family sizes of the household head those who did not participate on conservation agriculture deviates from its mean by 2.229. However, the maximum family sizes of respondent household those who participated on conservation agriculture was 16 while the minimum was 2. The standard deviation of family size of those farmers who participated on conservation agriculture was 2.191.

Table 5 Summary statistics of continuous variables and their mean difference test used binary logit model (n=142)

List of variables	Total respondent		Not participated[9]		Participated[10]		t -Value
	Mean	St. d	Mean	St. d	Mean	St. d	
Age	44.493	14.173	38.21	12.546	48.58	13.729	4.547***
Farm_size	1.352	0.949	1.3318	1.09880	1.3645	.84316	0.200
Experience	26.718	13.186	20.68	11.246	30.65	12.919	4.726***
Family_Labor	4.077	2.070	2.70	1.043	4.98	2.081	7.597***
Family_Size	5.831	2.275	5.11	2.229	6.30	2.191	3.155***
Extension_service_promoters	2.042	2.788	1.7500	2.89357	2.2326	2.71672	1.008

***, ** and * significant at 1%, 5% and 10% respectively

Source: Own Survey, 2013

---

[9] Farmers who did not participate on conservation agriculture
[10] Farmers who did participate on conservation agriculture

The highest level of education attained by respondent household who did not participate on conservation agriculture was certificate while the highest level of education attained by household head who participated on conservation agriculture was grade 11-12. The standard deviation of education level of household who participated on conservation agriculture was 1.010 while it was 1.05 for those farmers who did not participate on conservation agriculture. Out of all households who participated on conservation agriculture 50 percent (43 out of 86) were those who attended grade 1-8 while 64.29 percent (36 out of 56) of all who did not participate on conservation agriculture were those who were illiterate (Appendix a). According to the result of the household survey conducted from all respondents 86 were participated on conservation agriculture while 56 respondents were those who did not participated on conservation agriculture.

Table 6 Summary statistics of dummy and categorical variables used binary logit model (n=142)

List of variables	Total respondent		Not participated[11]		Participated[12]		χ^2 -Value
	Mean	St. d	Mean	St. d	Mean	St. d	
Education	1.289	1.140	0.66	1.049	1.70	1.007	37.113***
Sex	0.873	0.334	0.79	0.414	0.93	0.256	6.399**
Marital	0.859	0.349	0.768	0.426	0.918	0.275	10.317
Employment	0.852	0.356	0.66	0.478	0.98	0.152	6.396**
Topography	0.521	0.501	0.589	0.496	0.477	0.502	1.721
Membership	0.739	0.440	0.66	0.478	0.79	0.409	2.974*

***, ** and * significant at 1%, 5% and 10% respectively

Source: Own Survey, 2013

3.1. Conservation Agriculture as Adaptation Strategy to Climate Change

Conservation Agriculture can increase the ability of smallholder farmers to adapt to climate change by reducing vulnerability to drought and enriching the local natural resource base on which farm productivity depends. Conservation Agriculture aims at increasing the annual input of fresh organic matter, controlling soil organic material losses through soil erosion, and reducing the rate of soil organic material mineralization (Carlton and Antonio, 2012).

---

[11] Farmers who did not participate on conservation agriculture
[12] Farmers who did participate on conservation agriculture

Out of the total 142 respondents 130 were those households who perceive conservation agriculture as an adaptation strategy to climate change. 64 out of 130 households perceived conservation agriculture as an adaptation strategy were those whose average topography of their plots is flat while the rest 66 were those whose average topography of their plots is gentle, steep slope and mountainous. As illustrated on the following graph about 55 percent of the respondent households adopt the crop rotation technique of conservation agriculture. Cover crops and mulching was undertaken by 37 percent of total household respondent while minimum tillage and direct planting was undertaken by about 8 percent of sample households.

Figure 1 Households undertaking Conservation Agriculture Technique

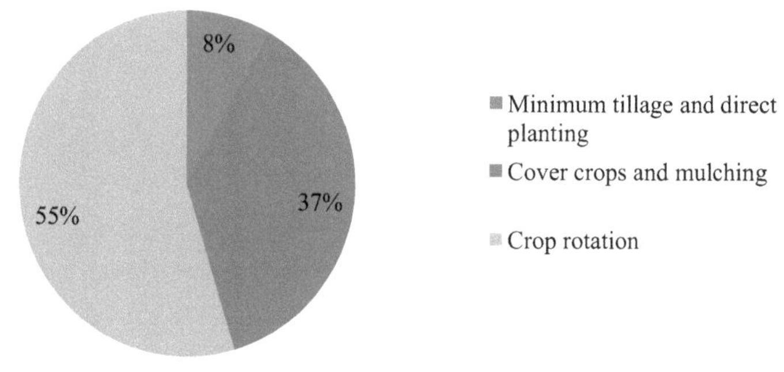

Source: Own Survey, 2013

3.2.   Results of the Econometric Model

3.2.1.   Farmers' Perception and Adaptation to Climate Change

Farmers should be able to adapt in order to reduce the negative impact of climate change in order to increase production and productivity. Adaptation to climate change is a two-step process which requires that farmers perceive climate change in the first step and respond to changes in the second step through adaptation. To get information on their perceptions of climate change, farmers were asked if they have observed any change in temperature or the amount of rainfall over the past years. The analysis of farmers' perceptions of climate change indicates that most of the farmers in this study are aware of the fact that temperature is increasing and the level of precipitation is declining (Appendix c). Different socio-economic and environmental factors affect the abilities to perceive and adapt to climate change.

Among the explanatory variables used in the model, 7 variables were significant with respect to outcome equation with less than 10 percent of the probability level while 5 variables were significant with respect to selection equation. The variables having a significant effect on adaptation to climate change in the study area are discussed below.

The number of relatives is one of the social capitals which increase the awareness of the households on their environment. As expected, households' number of relatives in development group was positively related with perception of climate change. One increase in number of relative of household head raises the probability of perceiving climate change by 0.16 percent.

Having access to farmer-to-farmer extension increases the likelihood of perceiving occurrence of climate change by 50.79 percent. Information on temperature and rainfall has a significant and positive impact on the likelihood perceiving climate change and access to information on climate change increases the probability of perceiving the occurrence of change in climate by 18.42 percent. Access to climate change Information is an important precondition for farmers to take up adaptation measures (Madison, 2006)

The agro-ecological setting of farmers influences the perception of farmers to climate change. As expected, different farmers living in different agro-ecological settings perceive the occurrence of climate change differently. The result of this study shows as one moves from Kolla to Woina dega local agro-ecology the probability of perceiving the occurrence of climate change decreases by 16.19 percent. Contrary to Deressa *et al.*, (2011) farmers living in Kolla (lowland) perceived more change in climate than farmers in Woina dega (mid-land) or Dega (high land).

The farmers' perception of change in duration of season significantly affects perception of climate change. As an individual farmer observe change in duration of season, his/her probability of perceiving change in climate increases by 3.44 percent. District dummy variable negatively and significantly affected perception of climate change. This shows that respondent households in Guto Gida district perceived the occurrence of climate change than those in Sasiga district.

Level of education of household took the expected sign and its coefficient was significant at less than 10 percent probability level. It had a positive and strong relationship with the dependent variable showing that literate household heads were more probability to adapt climate change on average. One level increase in education raises the probability of adaptation to climate change by 3.75 percent. This result is in line with Ayanwuyi *et al.,* (2010) who reported that education level of households had positive and significant relationship with perception of climate change.

Nonfarm income increases the probability of adapting the climate change. One birr increase in household nonfarm income leads to the increment of the probability of adaptation to climate change by 0.001 percent. This implies that households with income may get capital, land and labour. These factors serve as important factors for coping with adaptation (Apata *et al.,* 2009). So, adaptation to climate change depends on availability of income.

Livestock ownership is a sign of wealth to farmers (Sofoluwe *et al.,* 2011). The ownership of livestock is also positively related to the adaptation of climate change. An increase in access to livestock ownership raises the probability of adaptation to change in climate by 19.01 percent.

As expected, access to crop and livestock extension has a positive and significant impact on adaptation to climate change. Having access to crop and livestock production increases the probability of adapting climate change by 20.03 percent. This result is in line with Ayanwuyi *et al.,* (2010) who reported that access to extension facilities of households had positive and significant relationship with perception and adaptation of climate change.

Resource availability is generally expected to positively influence farmers' to adapt climate change. Hence, access to credit is expected to have positive relationship with farmers' adaptation to change in climate. Credit availability is one factor that leads household to adapt climate change. An increase in access to credit raises the probability of adaptation to climate change by 12.78 percent. Similar with this finding Charles and Rashid (2007) and Apata *et al.,* (2009) showed farmers with access to credit have higher chances of adapting to changing climatic conditions. This result is also in line with Ayanwuyi *et al.,* (2010) who reported that access to credit facilities of households had positive and significant relationship with perception of climate change and adaptation options.

Increment in temperature and adaptation to climate change were hypothesized to be related positively. As expected the result of this study shows the direct relationship between adaptation to climate change and perception of increase in temperature. Perceived change in temperature has significant effect in the likelihood of employing climate change adaptation strategies (ACCCA, 2010). For increment in perception of increase in temperature raises the probability of adaptation to climate change by 42.28 percent. Adaptation to climate change and precipitation were negatively related as expected. Perceiving decrease in precipitation raises the probability of adapting climate change by 64.76 percent.

Table 7 Result of Heckman two stage sample selection model (n=142)

Farmers' Perception to Climate Change (Selection equation)			Farmers' Adaptation to Climate Change (Outcome equation)		
	Marginal Effect			Marginal Effect	
List of variables	dy/dx	P-value	List of variables	dy/dx	P-value
Education	0.00349	0.180	Education	0.03757	0.040**
Sex	0.04491	0.265	Family_Size	0.01327	0.152
No_of_relatives	0.00160	0.015**	Sex	0.00286	0.965
Farm income	0.00000	0.802	Non-Farm Income	0.00001	0.030**
Local agroeco	-0.16196	0.002***	Livestock_Ownership	0.19014	0.001***
Information	0.18417	0.000***	Extension_on_Crop	0.20033	0.000***
Season_change	0.03442	0.002***	Credit	0.12781	0.001***
Farmer_extension	0.50799	0.009***	Farm_Size	-0.01752	0.337
Distance_ Input				0.00153	0.580
Distance_ Output				0.00153	0.431
Increase_Temperature				0.42288	0.000***
Decrease_Temperature				0.13554	0.190
Decrease_precipitation				0.64763	0.000***
District	-0.10635	0.036**		-0.08341	0.233

***, ** and * significant at 1%, 5% and 10% level respectively.

Source: Computed from own survey (2013)

3.2.2. Farmers' Participation in Conservation Agriculture

Before running the binary logit model all the hypothesized explanatory variables were checked for the existence of multicollinearity problem. The strong linear dependence might be the source of collinearity within independent variables. VIF (variance inflation factor) and correlation matrix was used for testing the association between the hypothesized variables. As per the appendix e variables with high value of VIF were age (52.01), experience (33.48), sex (18.71), marital status (18.69), and family size (12.73). The VIF values larger than 10 shows evidence of multicollinearity. From correlation matrix as shown on appendix f variables specified above were with collinearity problem.

Based on these tests from all the explanatory variables planned to be included in binary logit model age, experience, sex, marital status and family size were rejected from the regression. The overall VIF test for all independent variables planned to be included to binary logit model was 14.23, while the value after the regression was 3.67, showing the problem of multicollinearity was solved (Appendix e).   In solving the problem of heteroskedasticity literatures used robust standard errors (Charles and Rashid, 2007). To address the possibilities of heteroskedasticity in the model, the researcher estimated a robust model that computes a robust variance estimator based on a variable list of equation.

Finally, all hypothesized explanatory variables expect those with multicollinearity problem, were included in the binary logistic analysis. These variables were selected on the basis of available literature and the results of the survey studies. To determine the best subset of explanatory variables that are good predictors of the dependent variable, the binary logistic regressions were estimated, which is available in stata (version 10).

The binary logit model results used to study factors influencing the farmer's participation on conservation agriculture are shown in Table .8. The model explained about 47.09 percent of the total variation in the sample for participation on conservation agriculture. From the result of classification table 81.69 percent of the values were specified correctly (Appendix g). This shows observations were reasonably classified. The result of Wald test shows all variables included in the model were jointly significant since the value of χ^2 (51.58) is significant at 1 percent probability level (Appendix h). Among the explanatory variables used in the model, three variables were significant with respect to participation on conservation agriculture with less than 10 percent of the probability level. The significant explanatory variables on participation in study area are discussed below.

Education is expected to reflect acquired knowledge of environmental necessity. Education has positive impacts on participation on conservation agriculture and was significant at 1 percent level. Consistent with this expectation, binary logistic regression showed educational status of farmers to have a strong power in explaining participation on conservation agriculture. Holding other regressors constant, a change in household head education level by one unit, say one level, will increase the odds of being participated on conservation agriculture by the factor of 0.1542.

The possible justification for this finding was that educated farmers tend to conserve their environment, use agricultural extension services and adapt climate change than the illiterates. These are important instruments in boosting production which makes farmers to be wealthier and reverse the environmental problem (Table 8). This result is similar to findings by Fapojuwo *et al.*, (2010) which identified the higher the educational level of the farmer, the higher the tendency of using improved soil conservation techniques. Paulos (2002) identified that literate household heads were more opt to recognize the advantages of soil conservation and were willing to take part in it which is in line with the study.

Households' main employment was significant at 1 percent. The estimated coefficient for dummy variable main employment of household with the odds of being participator in conservation agriculture over non participator was positively correlated. This suggests that the probability of being participator on conservation agriculture increases if one has participated on on-farm employment, other factors being constant. This meant that farmers with on farm employment were more likely to participate on the conservation agriculture practices than those off farm. This is agreeing with the hypothesized idea which says off-farm employee may not participate on conservation agriculture because he/she may not think about environment since his/her income may not directly related to production of crops.

Households with larger number of economically active labor are supposed to be better in conservation agriculture practices, since they are less likely to have shortage of labor which is required to do conservation activities. The coefficient of active family labour was positive and significant at 1 percent probability level. A unit increase in active family labour increased the log-odds of participating on conservation agriculture by 0.2063 when the other variables are held constant (Table 8). Hence, households with more active family labour were better placed to participate on conservation agriculture than those with less active family labour. This might be so because of the practices of conservation agriculture are labour intensive since it requires application of conservation techniques.

Table 8 Binary logistic regression for conservation agriculture (142)

List of variables	dy/dx	P-value	Odds ratio	p-value
Education	0.1542***	0.008	2.2439***	0.003
Farm_Size	-0.0009	0.985	0.9952	0.985
Family_Labor	0.2063***	0.000	2.9489***	0.000
Employment	0.4945***	0.009	9.0995**	0.014
Topography	0.0028	0.978	1.0151	0.978
Extension_Service_Promoters	0.0273	0.176	1.1539	0.169
Membership	-0.1254	0.231	0.4867	0.298
District	0.0858	0.448	1.5672	0.446

Log likelihood = -50.389404   Wald $\chi^2(8)$ = 51.58 Prob > χ^2 = 0.0000 Pseudo R^2 = 0.4709

***, **, and * significant at 1%, 5% and 10% level respectively.

Source: Computed from own survey

3.3.   Climate Change Adaptation Measures and Causes of Non Adaptation

From all respondents who adapted the change in climate majority of them (45.33 percent) adapted the change through taking soil conservation measures. About 9.33 percent have taken measures of planting crop varieties in order to cope up climate change problem. 32 percent of them were participated on planting trees in reducing the problem caused by climate change. The remaining 13.34 percent of total households who adapted the change in climate were participated in irrigation activities in order to solve problem faced them through climate change.

Figure 2 Climate adaptation measures practiced

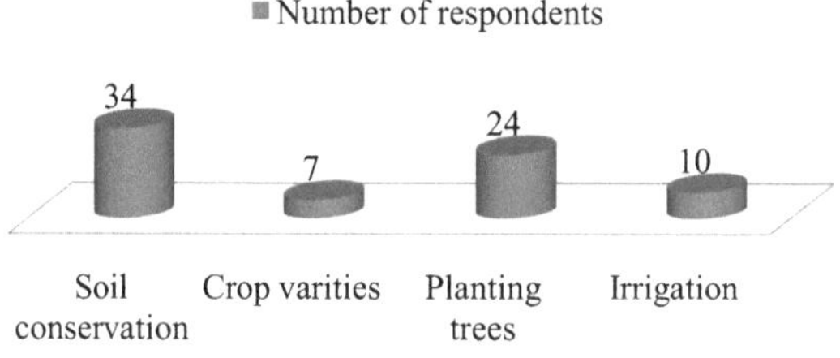

Source: Own Survey, 2013

Out of all respondent households who perceived the change in climate, those who did not adapt the change was not free of cause. Majority of them (47 percent) did not adapt because of shortage of labor while 32 percent of them did not adapt the change in climate because of lack of money to undertake the adaptation measures. From all those who did not adapt the change in climate 12 percent of them were absent from adapting the change because of lack of information and poor potential for irrigation.

Figure 3 Causes of non adaptation to climate change

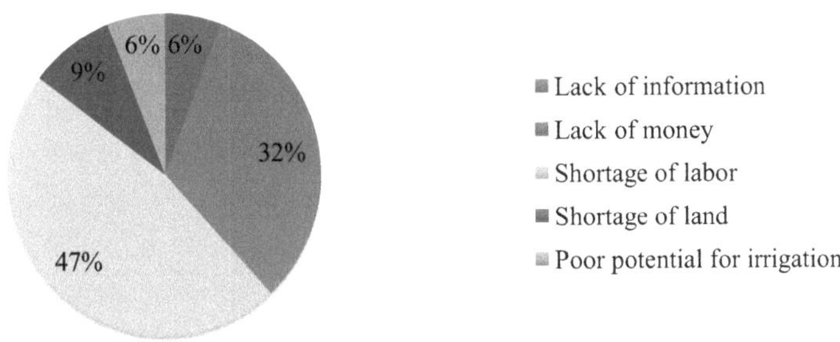

Source: Own Survey, 2013

4. Conclusion

The number of relatives is one of the social capitals which increase the awareness of the households on their environment. As expected, households' number of relatives in development group was positively related with perception of climate change. One increase in number of relative of household head raises the probability of perceiving climate change by 0.16 percent.

The agro-ecological setting of farmers influences the perception of farmers to climate change. As expected, different farmers living in different agro-ecological settings perceive the occurrence of climate change differently. The result of this study shows as one moves from Kolla to Woina dega local agro-ecology the probability of perceiving the occurrence of climate change decreases by 16.19 percent.

Level of education of household took the expected sign and its coefficient was significant at less than 5 percent probability level. It had a positive and strong relationship with the dependent variable showing that literate household heads were more probability adapt climate change. One level increase in education raises the probability of adaptation to climate change by 3.75 percent. Access to crop and livestock extension has a positive and significant impact on adaptation to climate change. Having access to crop and livestock production increases the probability of adapting climate change by 20.03 percent. This shows farmers with best access to crop and livestock extension adapt the impact of climate change more.

Resource availability is generally expected to positively influence farmers' to adapt climate change. Hence, access to credit is expected to have positive relationship with farmers' adaptation to change in climate. Credit availability is one factor that leads household to adapt climate change. An increase in access to credit raises the probability of adaptation to climate change by 12.78 percent.

Majority of respondent households perceive conservation agriculture as adaptation strategy to climate change. Out of the total 142 respondents 130 (91.55 percent) were those households who perceive conservation agriculture as an adaptation strategy to climate change. Active family labor and level of education of household significantly affect their participation in conservation agriculture at less than 1 percent probability level. Households with more active family labour were better placed to participate on conservation agriculture than those with less active family labour. Educated farmers tend to conserve their environment, use agricultural extension services and adapt climate change than the illiterates. The households main employment was significantly affect participation on CA at probability level less than 5 percent probability level. Farmers with on farm employment were more likely to participate on the conservation agriculture practices than those off farm.

5. References

Abera, B. (2003). Factors Influencing the Adoption of Introduced Soil Conservation Practices in Northwestern Ethiopia, discussion paper, Institute of Rural Development, University of Gottingen.

ACCCA. (2010). Improving decision-making capacity of small holder farmers in response to climate risk adaptation in three drought-prone districts of Tigray, northern Ethiopia, Mekelle, Ethiopia

Anemut, B., (2006). Determinants of Farmers' Willingness to Pay for the Conservation of National Parks: The Case of Simen Mountains National Park, MSc Thesis in Agricultural Economics, Haramaya University

Apata, T.G., K.D.Samuel, and A.O.Adeola., (2009). Analysis of Climate Change Perception and Adaptation among Arable Food Crop Farmers in South Western Nigeria, International Association of Agricultural Economists' 2009 Conference, Beijing, China

Bruce, A.M., Richard, M.A., and Brian, H. H., (2001). Global Climate Change and Its Impact on Agriculture

Carlton, P. and Antonio, A., (2012). Conservation Agriculture as a Strategy to Cope with Climate Change in Sub-Saharan Africa: The Case of Nampula, Mozambique

Ayanwuyi, E. Kuponiyi., F.A. Ogunlade., and Oyetoro, J., (2010). Farmers Perception of Impact of Climate Changes on Food Crop Production in Ogbomosho Agricultural Zone of Oyo State, Nigeria, *Global Journal of Human Social Science*, **10** (7): 33-39

Charles, N and Rashid, H., (2007). Micro-Level Analysis of Farmers' Adaptation to Climate Change in Southern Africa, IFPRI Discussion Paper 00714, August, 2007

Deressa, T.T., Hassan, R.M., and Ringler, C., (2011). Perception of and adaptation to climate change by farmers in the Nile basin of Ethiopia, *Journal of Agricultural Science*, **149**(2011): 23–31

Deressa, T.T., R. M. Hassan., Tekie, A., Mahmud, Y. and Claudia, R., (2008). Analyzing The Determinants of Farmers' Choice of Adaptation Methods and Perceptions of Climate Change in the Nile Basin of Ethiopia

Fapojuwo, O.E., Olawoye, J.E., and Fabusoro, E., (2010). Soil Conservation Techniques for Climate Change Adaptation among Arable Crop Farmers in Southwest Nigeria

Gujarati, D.N., (2004). Basic Econometrics, Fourth Edition, The McGraw–Hill Companies, 2004

Kothari., (2004). Research Methodology; Methods and Techniques, 2nd Revised Edition, New Age International Publishers, New Delhi, India

Maddison, D., (2006). The Perception of and Adaptation to Climate Change in Africa, CEEPA Discussion Paper No.10, Centre for Environmental Economics and Policy in Africa, University of Pretoria.

Paulos, A., (2002). Determinants of farmers' willingness to participate in soil conservation Practices in the highlands of Bale: the case of Dinsho farming system area, MSc Thesis in Agricultural Economics, Haramaya University

Shibru, T. and Kifle, L., (1998). Environmental Management in Ethiopia: Have the National Conservation Plans Worrked? Organization for Social Science Research in Eastern and Southern Africa (OSSREA), Addis Ababa, Ethiopia

Sofoluwe, N. A., Tijani, A. A. and Baruwa, O. I., (2011). Farmers' perception and adaptation to climate change in Osun State, Nigeria, *African Journal of Agricultural Research*, 6(20): 4789-4794

6. Appendix

Appendix a

Distribution of sample household head by level of education and their perception and adaptation to climate change and participation in conservation agriculture

Educational level of household	Farmers perception of occurrence of climate change			Farmers adaptation to climate change			Participation on Conservation agriculture		
	No	Yes	Total	No	Yes	Total	No	Yes	Total
Illiterate	20	32	52	20	12	32	36	16	52
Basic education	1	17	18	8	9	17	7	11	18
Grade 1-8	11	43	54	3	40	43	11	43	54
Grade 9-10	1	15	16	2	13	15	1	15	16
Grade 11-12	0	1	1	0	1	1	0	1	1
Certificate	0	1	1	1	0	1	1	0	1
Total	33	109	142	34	75	109	56	86	142

Source: Own Survey, 2013

Appendix b

Households Family Size and Farm Size with Adaptation to Climate Change

List of variables	Farmers who did not adapted climate change				Farmers who adapted climate change				Total Respondent			
	mean	max	min	sd	mean	max	min	sd	mean	max	min	sd
Total Family Size	4.882	9	2	1.805	6.773	16	3	2.147	6.183	16	2	2.220
Farm size in Hectares	1.234	3	0.2	0.818	1.537	7.25	0.1	1.073	1.442	7.25	0.1	1.007

Source: Computed from own survey

Appendix c

Farmers Perception of Change in temperature and precipitation

List of variables	Districts					
	Guto Gida		Sasiga		Total	
	No	Yes	No	Yes	No	Yes
Increase_Temperature	18	51	31	42	49	93
Decrease_Temperature	68	1	67	6	135	7
Nochange_Temperature	66	3	67	6	133	9
Decrease_Precipitation	28	41	24	49	52	90
Increase_Precipitation	60	9	71	2	131	11
Nochange_Precipitation	64	5	70	3	134	8

Source: Own Survey, 2013

Appendix d

Summary of variables included in the study

variable	Obs	Mean	Std. Dev.	Min	Max
age	142	44.49296	14.17307	22	90
education	142	1.288732	1.139746	0	5
family_size	142	5.830986	2.275281	2	16
sex	142	.8732394	.3338823	0	1
farm_size	142	1.351585	.9485677	.1	7.25
no_of_rela~s	142	7.415493	4.48677	0	24
experience	142	26.71831	13.18573	5	57
marital	142	.8591549	.349093	0	1
employment	142	.8521127	.3562449	0	1
membership	142	.7394366	.4404958	0	1
family_labor	142	4.077465	2.069961	0	13
information	142	.6830986	.4669156	0	1
farmer_ext~n	142	.7253521	.4479166	0	1
farm_income	142	7797.746	5695.917	1170	39960
nonfarm_in~e	142	4043.831	3813.691	0	27550
livestock_~p	142	.7605634	.4282502	0	1
extension_~p	142	.5985915	.4919185	0	1
credit	142	.4295775	.4967681	0	1
distance_i~t	142	15.84894	14.97311	.5	85
distance_o~t	142	13.4757	12.80922	.5	85
local_agro~o	142	.4507042	.4993253	0	1
increase_t~e	142	.6549296	.477074	0	1
decrease_t~e	142	.0492958	.2172512	0	1
nochange_t~e	142	.0633803	.2445082	0	1
decrease_p~n	142	.6338028	.4834696	0	1
increase_p~n	142	.0774648	.2682738	0	1
nochange_p~n	142	.056338	.2313895	0	1
extension_~s	142	2.042254	2.787696	0	12
topography	142	.5211268	.5013218	0	1

.

Source: Computed from own survey (2013)

Appendix e

VIF test conducted for variables planned to be included in binary logit model

vif, uncentered

Variable	VIF	1/VIF
age	52.01	0.019227
experience	33.48	0.029872
sex	18.71	0.053450
marital	18.69	0.053492
family_size	12.73	0.078524
family_labor	10.82	0.092391
employment	7.96	0.125674
membership	5.37	0.186102
farm_size	3.87	0.258397
education	3.01	0.332610
topography	2.17	0.460037
extension_~s	1.91	0.524666
Mean VIF	14.23	

VIF test conducted for variables included in binary logit model

Variable	VIF	1/VIF
employment	6.49	0.154085
family_labor	6.03	0.165769
membership	4.98	0.200895
farm_size	3.50	0.285671
education	2.50	0.400217
district	2.01	0.498454
extension_~s	1.98	0.504558
topography	1.90	0.525296
Mean VIF	3.67	

Appendix f

Correlation Matrix

corr participation_ca age education sex marital farm_size experience family_labor family_size employment topography extension_service_promoters membership district

	part~_ca	age	educat~n	sex	marital	farm_s~e	experi~e	family~r	family~e	employ~t	topogr~y	extens~s	member~p	district
particip~_ca	1.0000													
age	0.3587	1.0000												
education	0.4462	0.0754	1.0000											
sex	0.2123	0.0508	0.3019	1.0000										
marital	0.2118	0.0113	0.4060	0.7585	1.0000									
farm_size	0.0169	0.1746	0.0288	0.1014	0.0992	1.0000								
experience	0.3709	0.9365	0.0767	0.0804	0.0175	0.2206	1.0000							
family_labor	0.5403	0.3915	0.2580	0.0451	0.0839	-0.0347	0.4342	1.0000						
family_size	0.2576	0.2370	0.2733	0.1210	0.1395	0.1094	0.3109	0.6187	1.0000					
employment	0.4351	0.2477	0.1932	0.0798	0.1165	0.0788	0.2598	0.3138	0.1702	1.0000				
topography	-0.1101	-0.2151	0.0203	0.0161	0.0171	0.0743	-0.2255	-0.1759	-0.0404	-0.0817	1.0000			
extension_~s	0.0849	-0.1174	0.0140	0.1734	0.1884	0.2385	-0.0892	-0.1309	0.0928	0.1492	0.1415	1.0000		
membership	0.1447	0.2502	0.0379	0.1596	0.0825	0.3204	0.3365	0.3256	0.2813	0.2498	-0.1194	0.0841	1.0000	
district	-0.0349	0.1118	-0.1002	0.0107	0.0519	0.0493	0.1014	-0.1138	-0.1160	-0.1669	-0.1704	0.3039	-0.0956	1.0000

Appendix g

Classification Table

estat classification

```
Logistic model for participation_ca

                  ——————— True ———————
Classified |       D              ~D    |     Total
-----------+----------------------------+----------
     +     |       71             11    |       82
     -     |       15             45    |       60
-----------+----------------------------+----------
   Total   |       86             56    |      142

Classified + if predicted Pr(D) >= .5
True D defined as participation_ca != 0
------------------------------------------------------------
Sensitivity                     Pr( +| D)        82.56%
Specificity                     Pr( -|~D)        80.36%
Positive predictive value       Pr( D| +)        86.59%
Negative predictive value       Pr(~D| -)        75.00%
------------------------------------------------------------
False + rate for true ~D        Pr( +|~D)        19.64%
False - rate for true D         Pr( -| D)        17.44%
False + rate for classified +   Pr(~D| +)        13.41%
False - rate for classified -   Pr( D| -)        25.00%
------------------------------------------------------------
Correctly classified                             81.69%
```

Appendix h

Wald Test for Binary Logit Model

test education farm_size family_labor employment topography extension_service_promoters membership district

```
( 1)   education = 0
( 2)   farm_size = 0
( 3)   family_labor = 0
( 4)   employment = 0
( 5)   topography = 0
( 6)   extension_service_promoters = 0
( 7)   membership = 0
( 8)   district = 0

         chi2( 8) =   51.58
       Prob > chi2 =   0.0000
```

28

Appendix i

Heckman Two Stage Selection Model Stata Result

heckman climate_adaptation education family_size sex nonfarm_income livestock_ownership extension_on_crop credit farm_size distance_input distance_output increase_temperature decrease_temperature nochange_temperature decrease_precipitation increase_precipitation nochange_precipitation district, twostep select(climate_perception = education sex no_of_relatives farm_income local_agroeco information season_change farmer_extension district) rhosigma level(96) first

```
Iteration 0:   log likelihood = -76.985773
Iteration 1:   log likelihood = -31.895152
Iteration 2:   log likelihood = -23.052547
Iteration 3:   log likelihood = -19.370578
Iteration 4:   log likelihood = -18.020515
Iteration 5:   log likelihood = -17.753849
Iteration 6:   log likelihood = -17.734208
Iteration 7:   log likelihood = -17.734042
```

```
Probit regression                          Number of obs   =        142
                                           LR chi2(9)      =     118.50
                                           Prob > chi2     =     0.0000
Log likelihood = -17.734042                Pseudo R2       =     0.7696
```

climate_pe~n	Coef.	Std. Err.	z	P>|z|	[95% Conf.	Interval]
education	.2934354	.2188099	1.34	0.180	-.1354242	.722295
sex	1.123291	1.007359	1.12	0.265	-.8510965	3.097679
no_of_rela~s	.1345489	.0550501	2.44	0.015	.0266528	.2424451
farm_income	.0000173	.0000691	0.25	0.802	-.0001181	.0001527
local_agro~o	-3.029241	.9769763	-3.10	0.002	-4.944079	-1.114402
information	2.564369	.6787151	3.78	0.000	1.234111	3.894626
season_cha~e	1.765614	.5636205	3.13	0.002	.660938	2.87029
farmer_ext~n	3.681826	1.418177	2.60	0.009	.9022497	6.461402
district	-2.889799	1.377876	-2.10	0.036	-5.590386	-.1892125
_cons	-2.245949	1.076089	-2.09	0.037	-4.355045	-.1368529

```
Heckman selection model -- two-step estimates      Number of obs    =       142
(regression model with sample selection)           Censored obs     =        33
                                                   Uncensored obs   =       109

                                                   wald chi2(18)    =    570.20
                                                   Prob > chi2      =    0.0000
```

	Coef.	Std. Err.	z	P>|z|	[96% Conf.	Interval]
climate_ad~n						
education	.0375671	.0183252	2.05	0.040	-.0000682	.0752024
family_size	.0132721	.0092737	1.43	0.152	-.0057738	.032318
sex	.0028616	.0651669	0.04	0.965	-.1309749	.136698
nonfarm_in~e	.0000106	4.86e-06	2.18	0.030	5.93e-07	.0000206
livestock_~p	.1901363	.0566335	3.36	0.001	.0738252	.3064474
extension_~p	.2003349	.0493694	4.06	0.000	.0989426	.3017272
credit	.1278068	.0369474	3.46	0.001	.0519262	.2036874
farm_size	-.0175235	.0182665	-0.96	0.337	-.0550383	.0199913
distance_i~t	.0015267	.0027578	0.55	0.580	-.0041371	.0071905
distance_o~t	.001533	.0019467	0.79	0.431	-.0024651	.005531
increase_t~e	.4228809	.0774249	5.46	0.000	.2638697	.5818921
decrease_t~e	.1355429	.1033864	1.31	0.190	-.0767868	.3478725
decrease_p~n	.6476299	.0680313	9.52	0.000	.5079107	.787349
nochange_p~n	.1542435	.0919279	1.68	0.093	-.0345532	.3430403
district	-.0834073	.0699884	-1.19	0.233	-.227146	.0603314
_cons	-.735964	.1053212	-6.99	0.000	-.9522673	-.5196606
climate_pe~n						
education	.2934354	.2188099	1.34	0.180	-.1559453	.7428161
sex	1.123291	1.007359	1.12	0.265	-.9455716	3.192154
no_of_rela~s	.1345489	.0550501	2.44	0.015	.0214899	.2476079
farm_income	.0000173	.0000691	0.25	0.802	-.0001246	.0001592
local_agro~o	-3.029241	.9769763	-3.10	0.002	-5.035705	-1.022777
information	2.564369	.6787151	3.78	0.000	1.170458	3.958279
season_cha~e	1.765614	.5636205	3.13	0.002	.6080789	2.923149
farmer_ext~n	3.681826	1.418177	2.60	0.009	.7692461	6.594406
district	-2.889799	1.377876	-2.10	0.036	-5.71961	-.0599885
_cons	-2.245949	1.076089	-2.09	0.037	-4.455966	-.035932
mills						
lambda	.0393974	.0749142	0.53	0.599	-.1144575	.1932523
rho	0.22093					
sigma	.17832174					
lambda	.03939741	.0749142				

Appendix j

Stata Result of Binary Logit Model

logit   participation_ca   education   sex   farm_size   family_labor   employment   topography extension_service_promoters membership district, vce(robust) level(96) or

```
Iteration 0:   log pseudolikelihood = -95.23388
Iteration 1:   log pseudolikelihood = -57.653693
Iteration 2:   log pseudolikelihood = -51.675124
Iteration 3:   log pseudolikelihood = -50.478187
Iteration 4:   log pseudolikelihood = -50.390071
Iteration 5:   log pseudolikelihood = -50.389404
Iteration 6:   log pseudolikelihood = -50.389404


Logistic regression                          Number of obs   =        142
                                             Wald chi2(8)    =      51.58
                                             Prob > chi2     =     0.0000
Log pseudolikelihood = -50.389404            Pseudo R2       =     0.4709
```

particip~_ca	Odds Ratio	Robust Std. Err.	z	P>|z|	[96% Conf.	Interval]
education	2.243904	.6163761	2.94	0.003	1.27644	3.944648
farm_size	.9952056	.2477852	-0.02	0.985	.5968162	1.65953
family_labor	2.948923	.8094737	3.94	0.000	1.678147	5.181995
employment	9.099545	8.151097	2.47	0.014	1.44562	57.27766
topography	1.015141	.5522703	0.03	0.978	.3321133	3.102893
extension_~s	1.153944	.1201382	1.38	0.169	.9318048	1.429041
membership	.4867585	.3368092	-1.04	0.298	.1175297	2.01595
district	1.567239	.9247618	0.76	0.446	.4664914	5.265343

31